3rd Grade Science Volume 1

© 2013 Todd Deluca
OnBoard Academics, Inc
Newburyport, MA 01950

800-596-3175
www.onboardacademics.com

Table of Contents

Weather and Climate

The Three Main Ingredients of Weather.

Wind

Temperature

Precipitation

When we talk about the weather we are mostly talking about five particular things; wind, temperature, precipitation, humidity and pressure. Lets look at each of these things and how they affect weather.

The temperature is the measurement of how hot or cold it is. The higher the temperature the hotter it feels. Temperature is determined by how much sunlight we get. In the USA we use the fahrenheit scale to measure temperature. Seventy degrees means a nice warm day and 40 degrees means its quite chilly. Most people in the world use a different scale called celsius in the celsius scale it would be nice warm day if the temperature were about 20 degrees.

Wind is the movement of air and its an important part of weather. The condition of the wind is often used to describe weather by wind by saying its breezy or calm outside. Winds also can affect the weather by bringing new weather conditions with them. For example when winds blow in from the ocean they often bring rain. Wind also affects how we feel the temperature. Have you noticed that it feels colder in the winter when its windy outside. We call this the wind chill.

Precipitation is the term that describes when water falls outside in either liquid or frozen form. To put it simply, this is the term we give to rain, sleet, snow and hail.

Humidity is the amount of water vapor in the air. If its a hot sticky day then there is a lot of water vapor in the air.

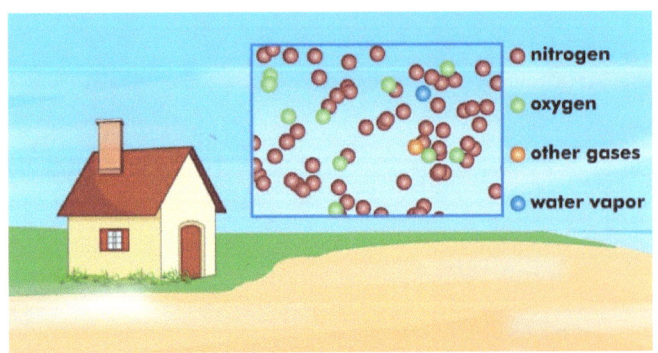

You may also hear the terms high and low pressure related to weather. This is simply the amount of press that air exerts in a downward direction. Its the weight of the air on top of you. Warm air is lighter so it exerts less pressure, cool air is heavier and exerts higher pressure.

We generally refer to weather in terms of five main characteristics: temperature, wind, precipitation, humidity and pressure. Climate refers to the type of weather that a region generally experiences.

www.onboardacademics.com

5 Major Types of Climates

A region's **climate** describes the type of weather that it typically experiences. There are five main types of climate: tropical, dry, temperate, cold, and polar. A region's climate depends on many factors, including its distance from the equator, the elevation of the region, how close the region is to the ocean, as well as particular ocean currents.

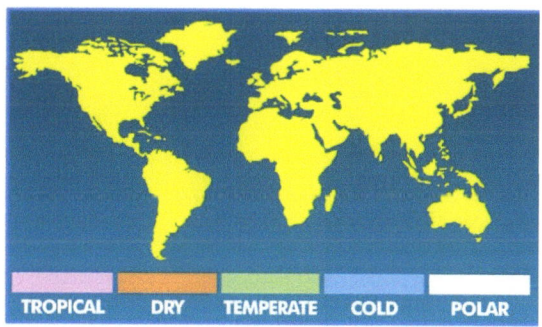

TROPICAL   DRY   TEMPERATE   COLD   POLAR

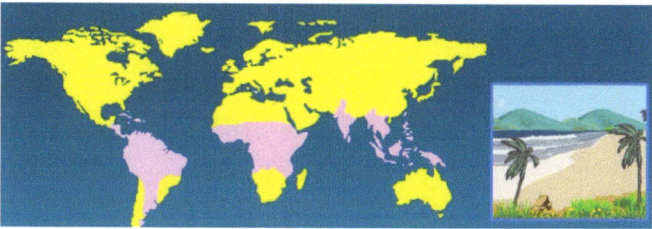

Tropical Climates have high temperatures as well as large amounts of rain all year

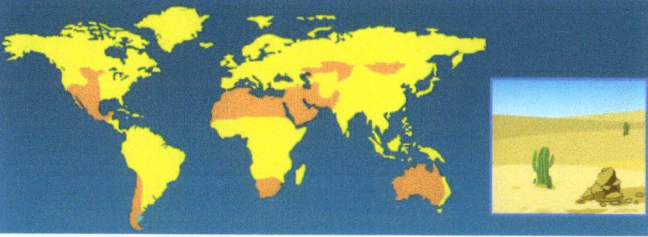

Dry climates have little rain during the year and experience a big temperature change during the day; very hot during the day and cold at night.

Temperate climates have dry warm summers and wet cool winters.

Cold climates are cold and have ice on the ground more then 3/4 of the year. The winters are severe and summers cool to warm.

Polar climates having freezing temperatures for most of the year. Ice and tundra are always present.

Label the following W for a description of weather and C for a description of climate.

1. It's raining outside. ☐

2. It's going to be sunny all week. ☐

3. Countries close to the equator are generally very hot. ☐

4. Seattle is often very wet. ☐

5. Deserts are hot and dry. ☐

6. Last year there was a huge thunderstorm on my birthday. ☐

5 Key factors that influence climate.

Latitude   Oceans   Topography

Elevation   Wind

Can you fill in the blanks?

_____ and large lakes tend to moderate a region's climate. Cold _____ currents can result in colder weather and more severe winters.

_____ describes how far a region is from the equator. _____ is regarded as the most important factor in a regions climate.

Global and local _____ such as winds that blow over an ocean and onto land can significantly affect the climate.

_____ describes how high a region is above sea level. _____ is a major factor in a region's climate.

_____ means the landforms in a region such as mountains which can significantly affect the climate.

See how elevation affects your climate.

The higher a location's elevation (or altitude) above sea level, the lower its temperature. On average, temperatures drop by about 6.5°C for every 1,000 m of extra elevation. This is because there is less air pressure at higher elevations. When air is less dense, it hold less heat.

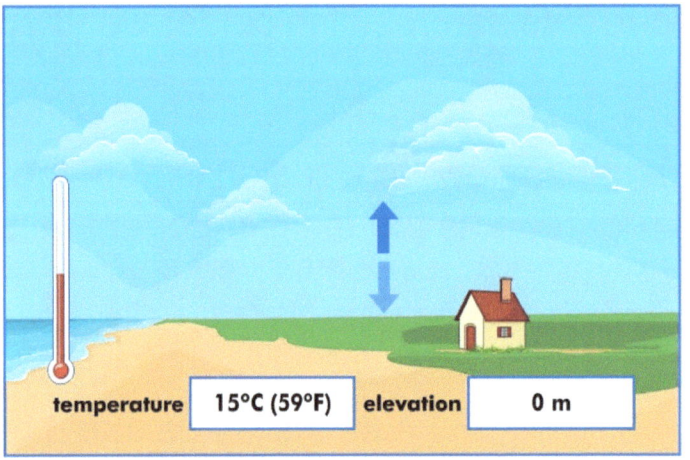

| temperature | 15°C (59°F) | elevation | 0 m |

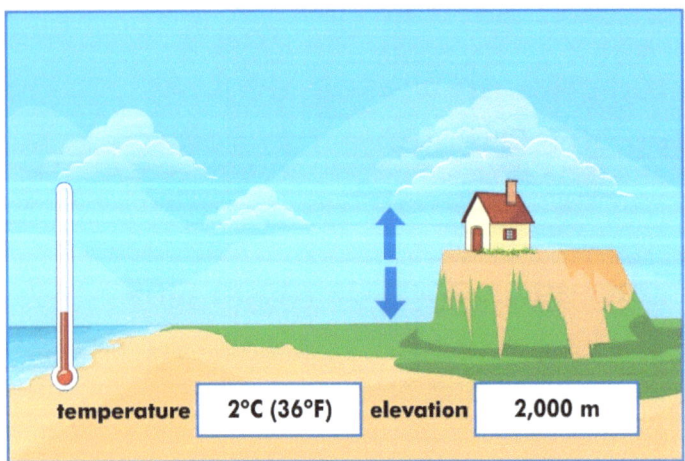

| temperature | 2°C (36°F) | elevation | 2,000 m |

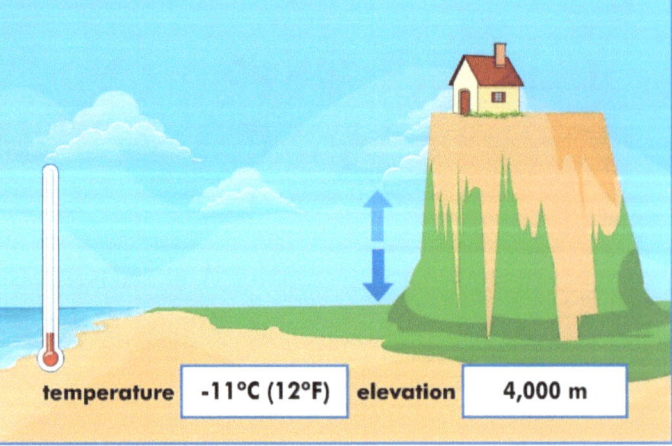

| temperature | -11°C (12°F) | elevation | 4,000 m |

Match the house on the position of the Earth with the proper latitude and temperature.

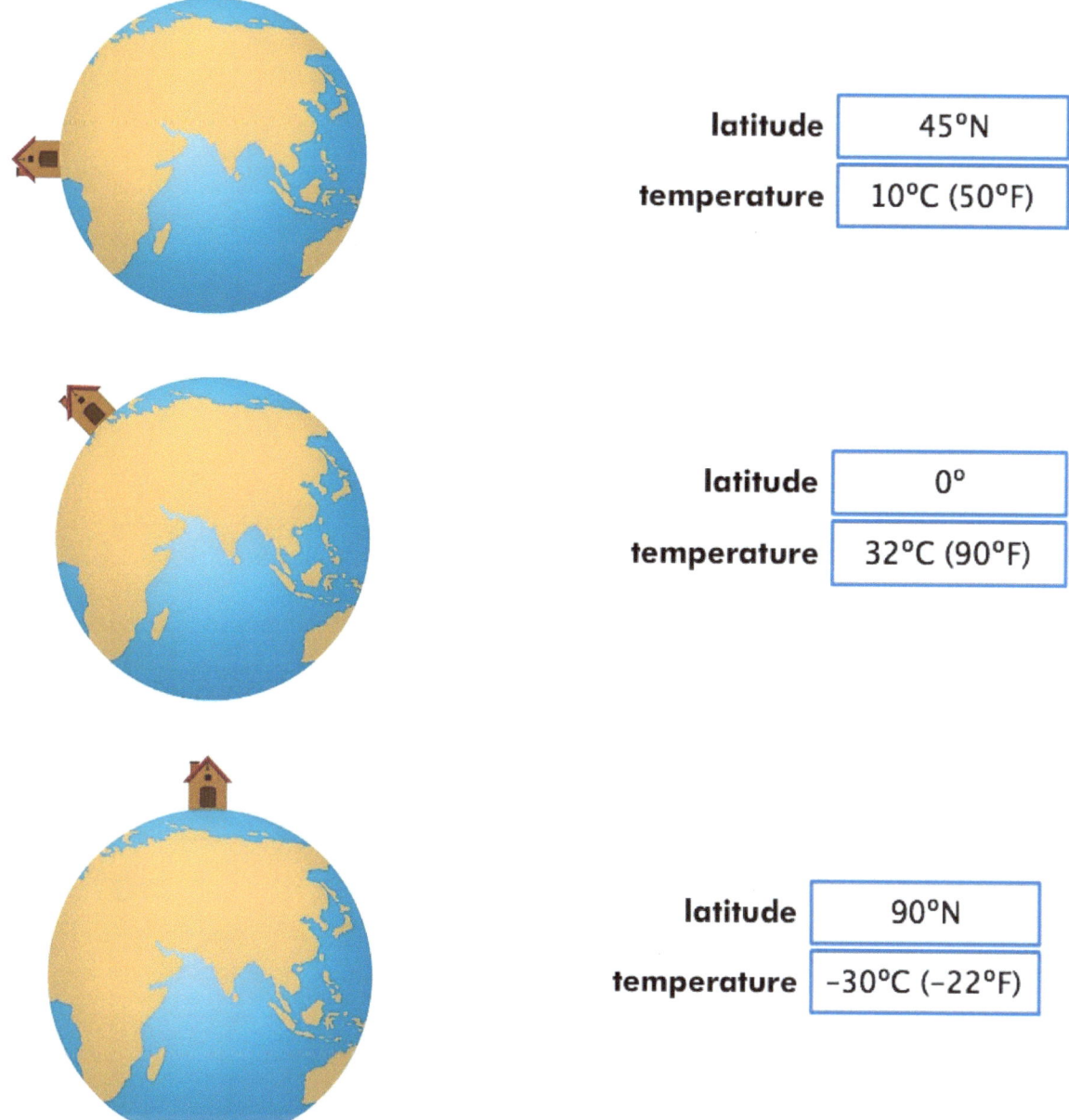

latitude	45°N
temperature	10°C (50°F)

latitude	0°
temperature	32°C (90°F)

latitude	90°N
temperature	−30°C (−22°F)

Locations closer to the equator get the most direct sunlight and so experience the warmest temperatures. Generally, the further a region is from the equator, the colder its climate.

What type of climates might these locations have?

warm	**cool**	**cold**	**hot**

How does the size of the mountain affect climate? _____

_____

_____

Elevation and Climate

Air that moves in from the ocean contains lots of water vapor. If the air is forced into a higher elevation because of a mountain, the water vapor starts to cool and then condenses to create rain clouds. Regions that are on the ocean side of mountains tend to have wet climates, while regions that are on the land side of mountains tend to have dry climates.

Water Temperature and Air Temperature

What happens to the air temperature when the water temperature is cold?

_____

What happens to the air temperature when the water temperature is warm?

_____

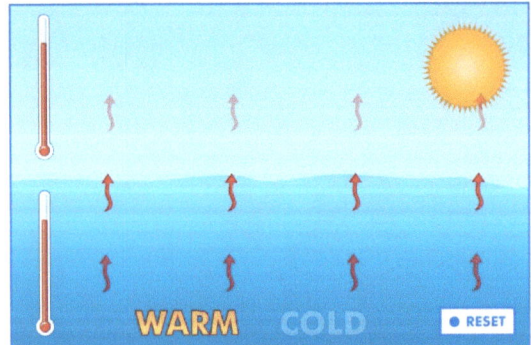 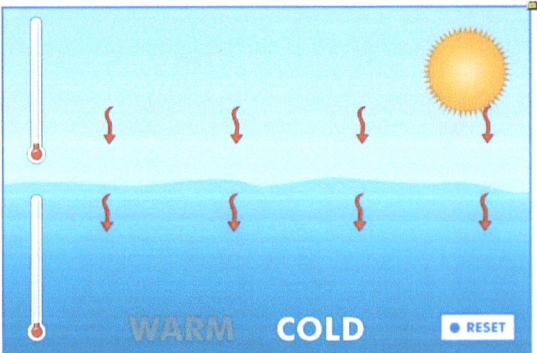

Newfoundland Canada has extremely cold and harsh winters but surprisingly Dublin Ireland, on the same latitude has moderate winters.

Study the illustration. Why do you think these two cities on the same latitude have such different winters?

_____

_____

_____

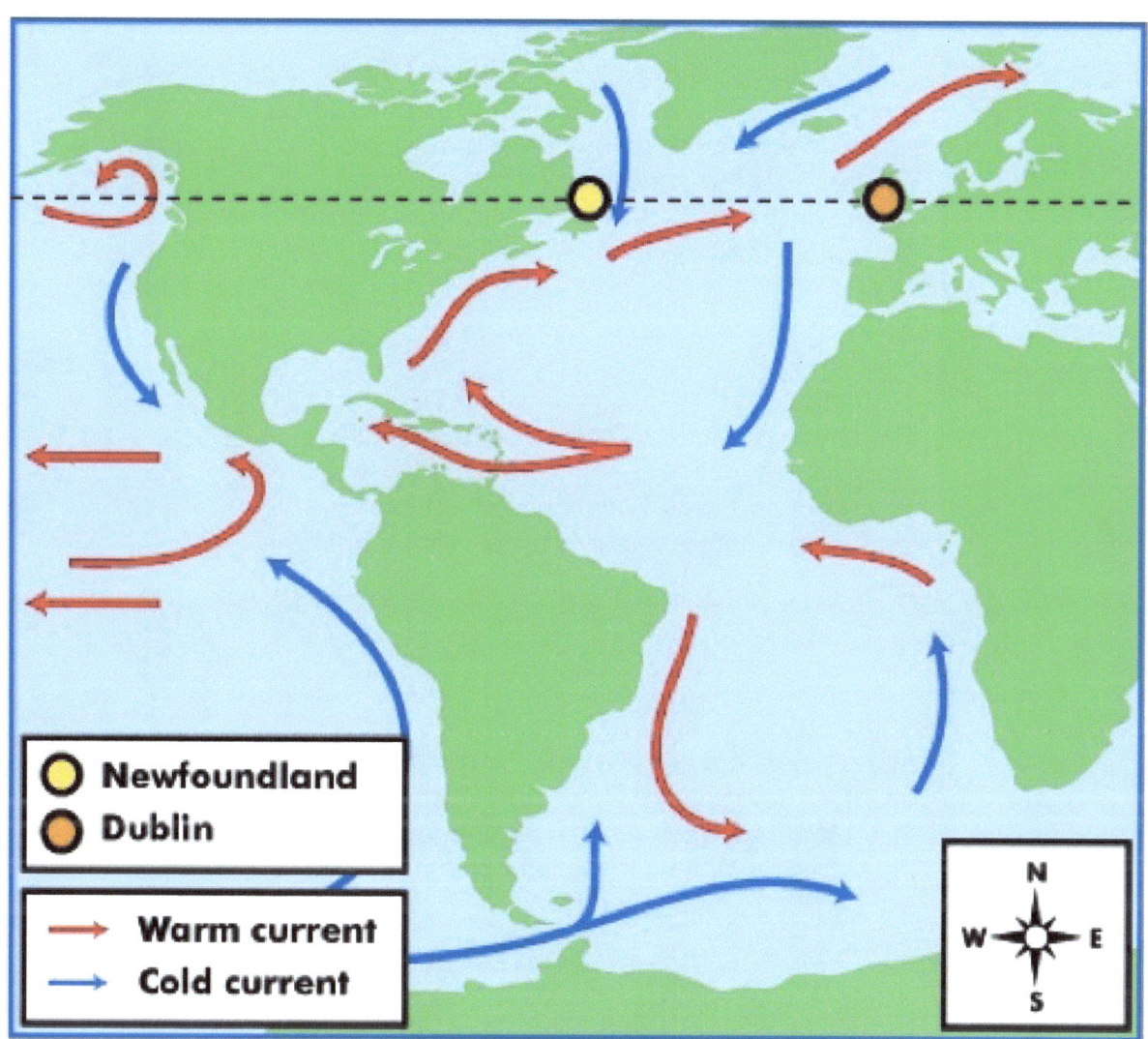

Wind affects climate.

Winds caused by warm air rising from the equator and moving toward the poles are know as the Prevailing Westerlies because they come from the west. Locations in the Prevailing Westerlies tend to have moderate rainfall throughout the year.

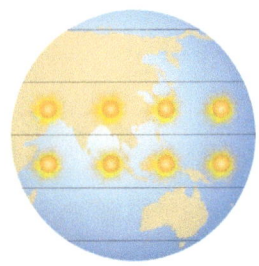

As warm air rises and moves toward the poles some of it cools and sinks back toward the equator. These are known as the Trade Winds, slow and steady winds that blow toward the equator from the northeast and the southeast. Locations in the Trade Winds tend to be dry throughout the year.

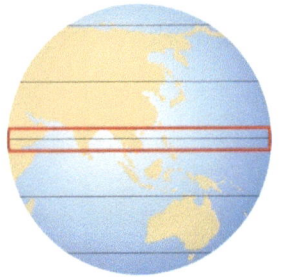

The Doldrums is the name given to the calm region just above and below the equator where the Trade Winds meet. Locations in the Doldrums tend to have large amounts of rain throughout the year..

At the poles, cold wind moves down toward the equator known as the Polar Easterlies because they blow from east to west. Locations in the Polar Easterlies tend to be very cold and have snow throughout the year.

Strong winds that accumulate sand and dust in very large .
A very windy snow storm.
A freezing rain storm that damages trees.

Extreme Weather

See if you can label the extreme weather descriptions correctly.

A very windy snow storm.

Strong winds that accumulate sand and dust in very large .

A freezing rain storm that damages trees.

A period of heavy rain and lightning.

These giant storms form over oceans with high winds and rain.

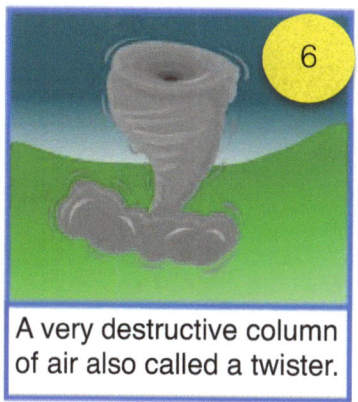

A very destructive column of air also called a twister.

ice storm  ◯   **thunderstorm** ◯   **dust storm** ◯

blizzard ◯   **hurricane** ◯   **tornado** ◯

Weather and Climate Quiz

1. Generally regarded as the most important factor affecting climate, I describe the distance of a location from the equator. What am I?

    _____

2. This ocean current helps some nations in Northern Europe experience relatively mild winters.

    _____

3. I am the term that is used to describe the altitude or height of a regions. _____

4. What is the name we give to the climate of a region that has wet cool winters and warm dry summers?

    _____

5. This is the name we give to the climate of a region that has high temperatures and lots of rain all year long.

    _____

6. I am the term that describes different types of landforms that can affect a region's climate.

    _____

Hurricanes

Is this a cyclone, a tropical cyclone, a hurricane or a typhoon?

Image: NASA

A hurricane, a typhoon and a cyclone are regional names for the same thing: a violent swirling storm that develops over warm ocean waters. Scientists have a single name to describe these storms: **a tropical cyclone.** However, we'll use the term *hurricane* in this lesson, since this is what we call tropical cyclones that occur in our part of the world.

Hurricanes

Hurricanes form over warm ocean waters near the equator. Warm moist air over the ocean rises to form thunder clouds.

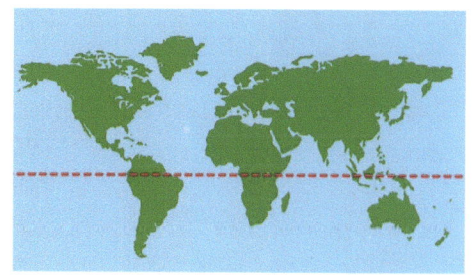

When the warm moist air rises there is less air near the surface of the ocean. We call this a low pressure area. Cool air from convergent winds (winds that blow from opposite directions) rush in to fill the low pressure area. This newly arrived cool air becomes warm and moist to form new thunder clouds. As the cycle continues the thunder clouds increase in number and begin to cluster.

 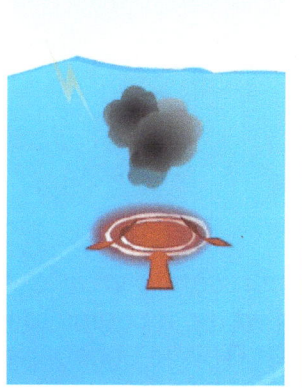 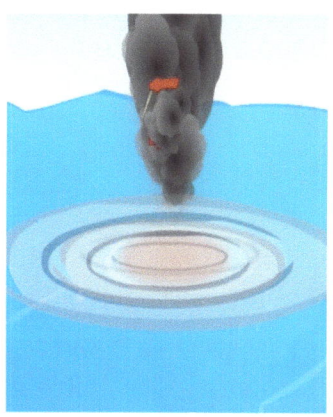 

At this stage the storm is classified as a Tropical Depression. The storm begins to spin counter clockwise as a result of the rotation of the Earth. When wind speeds increase to 63 kilometers or 39 miles per hour, it's classified as a tropical storm.

Air in the center of the storm begins to lose its moisture and this creates a calm area called the eye of the storm.

They eye of a storm is generally 30-60 kilometers or 20-40 miles wide and surrounded by the eye wall. This is a a thick wall of clouds with the heaviest rains and the strongest winds.

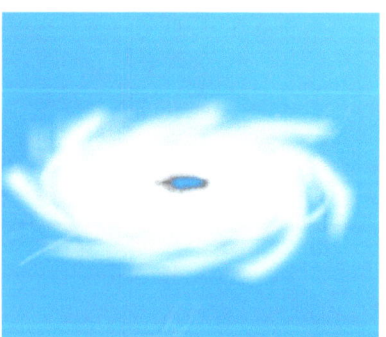

The growing storm gathers more pace and spins quickly fueled by the heat energy that is released with the water vapor rises and condenses. When winds reach 119 kilometers or 79 miles per hour the storm is classified as a tropical cyclone also known as a hurricane.

Hurricanes can last a week or longer and they can travel a great distance. When a hurricane moves over land they often have devastating consequences for life and property.

When a hurricane moves over land or passes over cool water, it begins to lose strength and slow down. This is because the storm will lose the warm moist air that acts as its energy.

In the USA, the peak hurricane threat is mid August until late October when the waters of the Atlantic Ocean and Gulf of Mexico have been warming over summer and the hurricanes have had time to develop.

Label the parts of a hurricane.

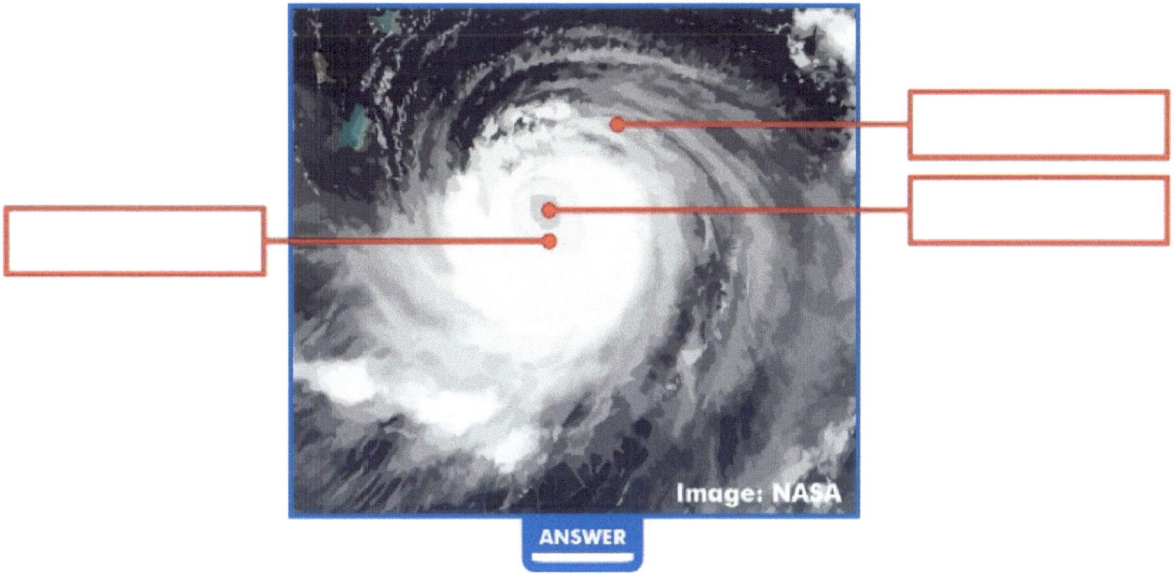

eye   rain bands   eyewall

The Saffir-Simpson Scale

Category	1
Wind Speed	74 - 95 mph
Damage	Some damage

Some damage to buildings, including broken windows and removal of shingles, danger of flying debris, and damage to power lines and trees.

Category	2
Wind Speed	96 - 110 mph
Damage	Moderate

Some damage to buildings, considerable damage to mobile homes, some trees blown down, and almost total loss of power.

Category	3
Wind Speed	111 - 130 mph
Damage	Extensive

Extensive damage to buildings, many mobile homes destroyed, large number of trees uprooted, high risk of injury from flying debris, and loss of power for several days.

Category	4
Wind Speed	131 - 155 mph
Damage	Extreme

Extensive damage to buildings, including building collapse, most mobile homes destroyed, large number of trees uprooted, very high risk of injury, and area uninhabitable for weeks.

Category	5
Wind Speed	>155 mph
Damage	Catastrophic

Many homes and almost all mobile homes destroyed, most trees uprooted, very high risk of injury, and area uninhabitable for months.

Where do most hurricanes form?

Indicate the are on the map where most hurricanes form.

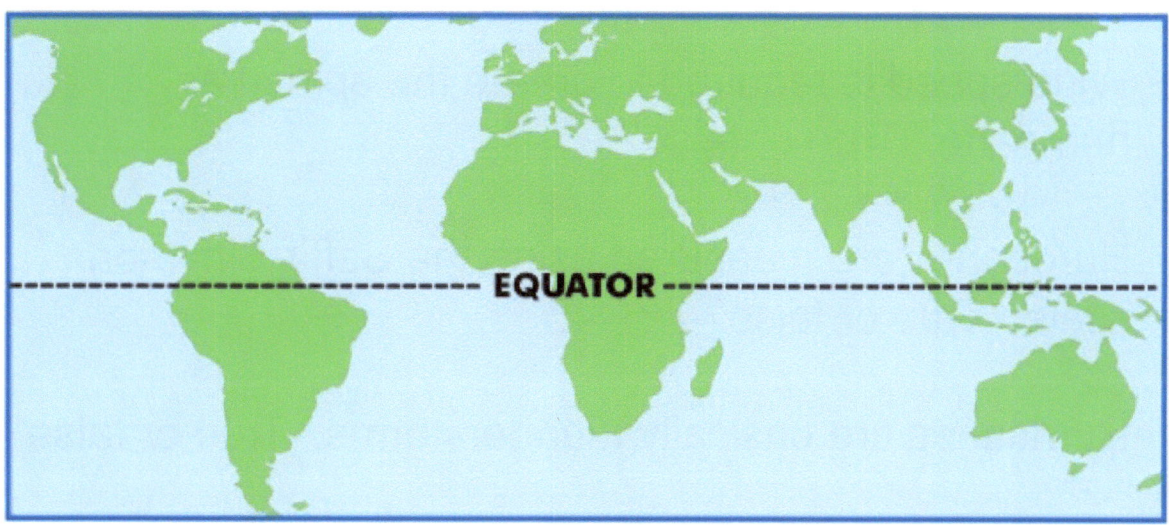

Most hurricanes occur in the tropics.
This is the name we give to the
band of Earth that is approximately
23° above and below the equator.

Hurricanes Quiz

1. Hurricanes form over _____.

2. Hurricanes occur on the equator. True or false?

3. Wind speed is required to create the spin in a hurricane. True or false?

4. Hurricanes are measured using the Saffir-Simpson scale. True or false?

5. Hurricanes are basically thunderstorms. True or false?

6. A combination of warm air from the ocean's surface and warm air form the storm leads to a hurricane. True or false?

7. What is the windiest and rainiest part of the hurricane? _____

Landslides

Would you like to live in this house? It's a nice house on a hillside overlooking the ocean.

And then there was a long, heavy rain storm……

Landslides can be destructive and deadly. Every year in the U.S., landslides kill about 25 people and cause damage estimated at more than a billion dollars. As we will discover in this lesson, too much water, sloping land, and loose soil can all be contributing factors to a landslide.

What is a landslide?

When debris, rocks or earth suddenly slide down a sloping surface; we call this a landslide. Landslides occur when the material on a sloping surface become loose and the Earth's gravity pulls it down the slope. Landslides can be very minor with just a few pieces of rock falling or or massive with large pieces of land sliding away.

There are three types of landslides.

Falls: A part of the rock which usually has a crack separates from a slope or cliff. Gravity then causes the rock to fall, bounce, or roll down the land.
Slides: Soil and rock debris slide downward on a slope.
Flows: Saturated soil and rocks flow down a slope, collecting debris as they move.

Here are three main causes of landslides.

water   **natural disasters**   **human activity**

Water
Water in the soil on a slope can be a good thing helping to bind things together. However, too much water is a key reason as to why landslides occur. Areas with heavy rainfall or melting snow causes the soil to be saturated. The wet heavy soil can trigger a landslide. The changes in water levels can also trigger a landslide. Fore example if a lake or river level decreases, the water that had held the sides of the river bed or lake bed is gone and no longer holding the sides in place.

Natural Disasters
Volcanoes and earthquakes can cause landslides. In both cases the ground becomes unstable and either explodes or shakes the earth loose triggering landslides.

Human Activity
Human activity such as construction, mining and logging and the use of heavy machinery and explosions in these industries are increasingly common causes of landslides. Farming can also cause landslides due to saturation of the earth due to excessive irrigation.

Label the action to prevent landslides with the appropriate illustration.

○ Plant vegetation in your area. The roots from trees and other plants will help to hold the soil in place.

○ Avoid soil saturation by setting up a proper drainage system.

○ Research the slope stability and landslide history in your area. Speak to a geologist before deciding where to build your home.

○ Set up barriers like seawalls to protect against land erosion.

○ Build homes with proper supports and materials set back from the edge of a slope.

Landslides Quiz

1. List the three main types of landslides.

    1)_____

    2)_____

    3)_____

2. Too much water can cause landslides to occur. True or false?

3. Name two types of natural disasters that can cause landslides.

    1) _____

    2) _____

4. Human activities also cause landslides. True or false?

5. When mud and soil mix with large amounts of water and come down a slope. This is a _____.

    a. slide

    b. flow

    c. fall

6. If you build in a landslide prone are, you should plant more trees. True or false?

Earth's History Through Rocks, Fossils and Tree Rings

Scientists estimate that the Earth is about 4,500,000,000 years old. But how do we know that?

Geologists are scientists who study Earth's history mostly by studying earths rocks and minerals. Geologists is a method called radioactive dating to calculate the age of Earth's rocks. This involves studying uranium, a silvery white chemical compound that is found in some metamorphic and igneous rock. Those are rocks that have ended up on the Earth's surface from volcanic activity. Uranium can be used to help date the rock because over a period of about 4 1/2 billion years half of the uranium atoms will decay and turn into lead. We call this is half-life. This means that when geologists study a rock that has uranium in it They can compare the amount of uranium in the rock with the amount of lead in the rock to calculate its approximate age. Using this technique scientists have calculated that the earth is about 4 1/2 billion years old. A similar technique is used to measure sedimentary rock except that geologists study the half life of a particular carbon atom found in sedimentary rock called Carbon 14.

Because the uranium that's found in some rocks slowly changes over time, scientists can use a process called radioactive dating to calculate the age of a rock.

Geologists can also date rocks by looking at rock layers.

When Owen's friends arrived for his party they all threw their coats on his bed. We can tell from the stack of coats when his friends arrived.

Layers of rock are formed in a similar way. When mineral sediments are deposited on existing rock new layers are formed. Over time as more sediment is deposited more and more layers form.

Scientists can date rock by observing its position in the layers since the newest layers are on top and the oldest layers are on the bottom. The layers also capture other information about changes tot he earth's surface over millions of years as well as changes to the Earth's plant and animal life.

How does an animal turn into a fossil.

Think about a fish swimming in the ocean millions of years ago. The fish is coming to the end of a long and happy fish life during which it has avoided the many fierce predators in the ocean at that time. When the fish finally dies of old age, the fish sinks peacefully to the bottom of the ocean.

Over time, bits and sand and rock accumulate on top of the fish and bury it below the ocean floor. Then, as the fish's body begins to rot, tiny bits of sand and rock replace the rotted body parts. This is one way that fossils are formed and why fossils sometimes look like animals that have been turned into stone.

The fossil's flat appearance is due to the weight of all the rock and sand above it. Fossils are sometimes discovered as a result of earthquakes or volcanoes which can force rocks up to the Earth's surface. Rocks that make it to the surface are great teachers if they contain fossils because they can show us what animals looked like millions or even billions of years ago. Sometimes the fossils offer amazing detail. This is really helpful because most of the animals in these fossils are extinct which means they're species is no longer living.

> **Fossils are found in rocks and are normally millions or even billions of years old. There are a number of different ways that animals can turn into fossils after they die.**

Use the fossils in this rock sample to answer questions about the history of the Earth.

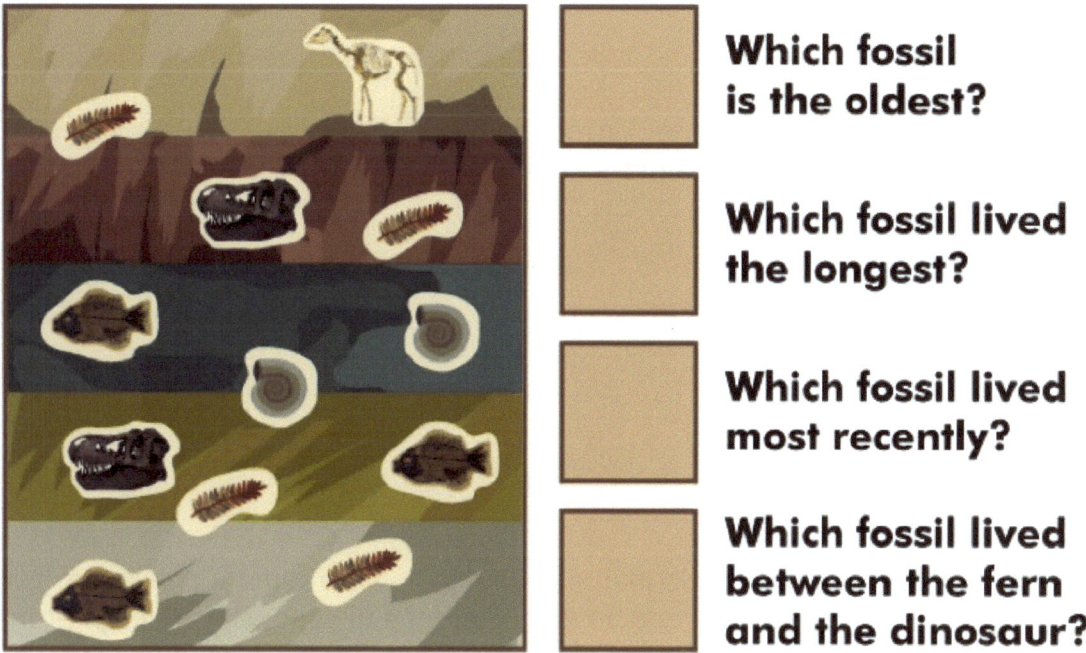

Which fossil
is the oldest?

Which fossil lived
the longest?

Which fossil lived
most recently?

Which fossil lived
between the fern
and the dinosaur?

Fossils of fish and other aquatic organisms discovered in mountains suggest to us that the rocks in the mountain used to be underwater. Similarly, if you found a tropical plant fossil in a more temperate (less hot) location, it would suggest that the climate of that location used to be much warmer. Fossils can tell us a lot about how the history of Earth has changed.

Tree Rings Tell Us About Earth's History

When a tree grows the width of its trunk increases to allow more water and nutrients to flow through it.

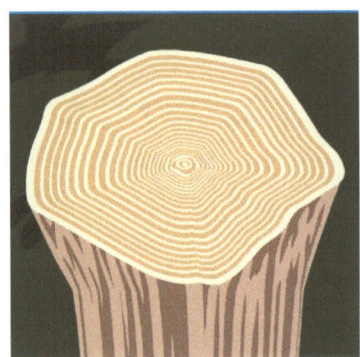

Trees that grow in a temperate climate grow more during the spring than any other time of year. Because spring wood is lighter in color that the wood that grows during the rest of the year, as the tree grows, visible rings appear inside the trunk. Each ring represents one year so by counting the rings we can determine the age of the tree. This is why tree rings are also called "annual growth rings."

Tree rings also reveal information about historical weather. Wide tree rings indicate years with plentiful rain while narrow tree rings indicate dry periods and droughts. Scientists study trees in temperate climates because trees in tropical climates have a more continual growing season and so have less pronounced rings.

How old are these trees?

11 years old   **20 years old**   **5 years old**

What do these tree rings tell us about historical weather?

Dry then wet climate

Wet then dry climate

Dry, then wet, then dry climate

Wet, then dry, then wet climate

Wet climate

Dry climate

Scientists can learn about the past climate by studying today's tree rings and the fossils of tree rings that are millions of years old.

Earth's History Through Rocks, Fossils and Tree Rings Quiz

1. How old is the earth? _____

2. Scientists who study the Earth's history through the study of rocks and minerals are called _____.
    archaeologists
    geologists
    paleontologists
    pulmonologists

3. Scientists can estimate the age of rocks by looking at rock layers. True or false?

4. Radioactive dating is the process used to calculate the age of _____.
    trees
    rivers
    rocks
    soil

5. Fossils cannot tell us how the Earth has changed over time. True or false?

Newburyport, MA 01950

1-800-596-3175

OnBoard Academics employs teachers to make lessons for teachers! We create and publish a wide range of aligned lessons in math, science and ELA for use on most EdTech devices including whiteboard, tablets, computers and pdfs for printing.

All of our lessons are aligned to the common core, the Next Generation Science Standards and all state standards.

If you like our products please visit our website for information on individual lessons, teachers licenses, building licenses, district licenses and subscriptions.

Thank you for using OnBoard Academic products.

www.ingramcontent.com/pod-product-compliance
Lightning Source LLC
Chambersburg PA
CBHW050839180526
45159CB00004B/1957

* 9 7 8 1 4 9 7 4 8 1 8 6 2 *